A Glossary of
French Bloodstock Terminology

A Glossary of French Bloodstock Terminology

MARY-LOUISE KEARNEY

J. A. ALLEN
LONDON

Published in Great Britain in 1981 by
J. A. ALLEN & COMPANY LTD.,
1, LOWER GROSVENOR PLACE,
LONDON SW1W 0EL

ISBN 0-85131-354-X

British Library Cataloguing in Publication Data.
Kearney, Mary-Louise
A Glossary of French Bloodstock Terminology.
1. Thoroughbred Horse — Dictionaries — French
2. French Language — Dictionaries — English
1. Title
636.1'32'03   SF 278
ISBN 0-85131-354-X

*The compiler and the publishers wish to
extend their grateful thanks to the Société
d'Encouragement and Messrs. Weatherby for
their kind assistance*

Printed and bound in Great Britain
at The Pitman Press, Bath

Contents

Preface 7

I Pedigree 9
    Breeds 9
    Description 10
    Pedigree 11
    Breeding 12

II Identification 15
    Head 15
    Neck 15
    Body 15
    Fore Limbs 17
    Hind Limbs 17
    Coat Colour 17
    Marks 18

III Commerce 19
    Sales 19
    Diseases 20
    Racing Documents 21
    Personnel 22

IV Racing Administration 24
    The Rules of Racing 24
    Racing Colours 27
    Race Categories 31
    The Organization of a Race 33
    The Going 37
    The Betting System 37

V Performance 39
    The Gaits 39
    Training Terms 39
    The Track Record 44

Preface

The history of the French turf is as colourful as that of its English counterpart. Both nations share a passion for the thoroughbred and for the spectacle of the racing industry.

Thoroughbred breeding in France dates from the 17th century when Louis XIV sent his special envoy, Garsault, to study the organization of the English studs. Under the patronage of Marie Antoinette, the turf became an institution in France and this tradition was continued by Napoleon, who appreciated the link between breeding and performance.

The 19th century marked the meteoric rise in popularity of the racing industry. In 1833, Lord Henry Seymour founded the *Société d'Encouragement pour l'amélioration des races de chevaux en France*, a circle of patrons dedicated to the promotion of breeding and racing. The French Stud Book was instituted in the same year and regular meetings were organized. A classic programme was established, including the first French Derby, le Prix du Jockey-Club, in 1836 and the Prix de Diane in 1843. The success of these ventures stimulated interest amongst enthusiasts in other disciplines. In 1863, the *Société des Steeple-Chases de France* was founded, while a racing authority for trotting, the *Société d'Encouragement à l'Elevage du cheval français*, was organized in 1864. From these orginis, there developed more than three hundred provincial clubs, which currently sponsor meetings all over France.

Today the French racing industry is the fourth largest in the world. The French champion has his home amongst the green, undulating pastures of Normandy, where the annual Deauville sales present the best crops of the French studs. During his career, the thoroughbred is protected by a variety of institutions, including the Stud Service of the Ministry of Agriculture, which works in partnership with each racing authority. Chantilly remains the principal centre of French training, producing winners which attract attention world wide. At home, racing provides a popular past time for millions of "turfistes", who play the Tiercé each week and throng the race courses.

Enthusiasts have remarked that the turf is a sport in England, an enterprise in the United States and an entertainment in France. Whatever the special features of each country, the racing industry has become an international fraternity, whose professionals work in close communication. This glossary aims to give a repertoire of the essential terms used in the French system today. It is hoped that it will give something of value to all racing enthusiasts so that they may be better acquainted with the French bloodstock industry.

I Pedigree

BREEDS

l'ambleur   *the pacer*
Pacing is forbidden on the French trotting scene, so this term
refers to the type of horse popular in the United States and
Australasia.

l'anglo-arabe   *the Anglo-Arab*
This term refers to any racing breed, other than the
thoroughbred

L'A.Q.P.S.A.   *the non-thoroughbred*
This term refers to any racing breed, other than the thoroughbred of the French Stud Book.

l'arabe   *the pure-bred Arab*
A magnificent selection of these horses stands at the national
Haras du Pompadour in the Limousin region of France.

le pur-sang anglais   *the English thoroughbred*
In racing parlance, this breed dominates the industry.

le cheval de selle français   *the French saddle horse*
This description applies to the horse descended from one pure-
blooded parent, usually of the thoroughbred, Arab or Anglo-
Arab breeds. There is a Stud Book for this hardy race, which is
particularly suited to a jumping career.

le cheval de haies   *the hurdler*
le hurdle-racer
le cheval d'obstacles   *the steeple-chaser*
le cheval de steeples
le steeple-chaser
An English influence is evident in the variety of these terms.

le trotteur français   *the French trotter*
Normandy is the cradle of this breed. In the 19th century,
Anglo-Normans were crossed with English thoroughbreds or
Norman horses. More recently, the American trotter has been
mated with the French variety to experiment with the race.

DESCRIPTION

le cheval cryptorchide   *the rig*
le cheval monorchide
Said of the horse retaining one testicle in the abdomen.
le cheval d'âge   *aged*
In France, this denotes horses 10 years old and above.
le cheval entier   *the horse*
In general, this means the horse which has not been gelded but according to the Rules of Racing, it specifically refers to the male aged 5 years or over.
l'étalon   *the stallion*
The male at stud.
le foal   *the foal*
This term describes the offspring from its birth until January 1st of the following year.
le hongre   *the gelding*
This denotes the gelded male at any age.
la jument   *the mare*
Said of the female aged 5 years or over, according to the Rules of Racing. Also used to denote the female at stud.
le poulain   *the colt*
The male until his 5th year.
la pouliche   *the filly*
The female until her 5th year or until she commences stud duties.
le poulain sevré   *the weanling*
le weanling
The weaned foal.
la poulinière   *the brood mare*
This term is more related to the breeding area, as it is used to designate any female which has been served at stud.
le yearling   *the yearling*
Said of the colt or filly from January 1st until December 31st of the year following its birth.

PEDIGREE

la branche   *the bloodline*

The direct male or female line of descent is also known by several other terms:

la lignée
le côté paternel
le côté maternel
le courant de sang
le rameau
la branche-maîtresse   *the tap-root strain*
le chef de race   *the foundation stallion*
l'étalon souche

These two terms denote the dominant male ancestor in a pedigree.

la deuxième mère   *the granddam*
la grand-mère
le deuxième père   *the grandsire*
le grand-père
la fille   *the daughter*
le fils   *the son*
le frère   *the brother*

This relationship includes the following:

le propre frère   *the full brother*
le demi-frère   *the half brother*
le frère utérin
le trois-quart frère   *the three-quarter brother*
la jument-base   *the tap-root mare*
la mère   *the dam*
le papier   *the pedigree*
le pedigree
le père   *the sire*

Other terms for the same title are:

l'auteur
le reproducteur
le sire
la petite-fille   *the granddaughter*
le petit-fils   *the grandson*

la production	*the progeny*

Useful to note are:

les premiers produits	*the first crop*
les derniers produits	*the last crop*
la soeur	*the sister*

Thus, the series:

la propre soeur	*the full sister*
la demi-soeur	*the half sister*
la soeur utérine	
la trois-quart soeur	*the three-quarter sister*
la souche maternelle	*the family*
la famille	*the tail line*
	the bottom line

BREEDING

The key aspects of the breeding cycle are designated by the following terms:

accepter l'étalon	*to accept the stallion*
admettre l'étalon	
l'allaitment	*the suck*
l'atavisme	*atavism*
le boute-en-train	*the teaser*
le souffleur	
la castration	*gelding*
les chaleurs	*heat*

Associated terms, used to describe the mare:

être chaude	*to be in season*
revenir en chaleur	*to break her service*
couler	*to slip*
le cycle oestral	*the oestrous cycle*
donner	*to throw*
engendrer	
produire	
l'effectif	*the string*
passer une jument à la barre	*to tease the mare*
la qualité héréditaire	*prepotency*

la fécondité (mares)	*fertility*
la fertilité (stallions)	
faire la monte	*to stand at stud*
la gestation	*pregnancy*
l'inbreeding	*inbreeding*
la consanguinité	
une génération libre de consanguinité	*a free generation*
être inbred sur	*to be inbred to*
l'insémination artificielle	*artificial insemination*
la jument de courses	*the race mare*
la jument pleine	*the in foal mare*
la jument suitée	*the mare with foal at foot*
être liste pleine	*to have a full book of mares*
être mal né	*to be chance bred*
la maiden	*the maiden*
la monte	*covering*

Hence:

la monte en liberté	*covering in liberty*
la monte en mains	*hand covering*
la saison de monte	*the covering season*
l'outcrossing	*outcrossing*

Associated terms are:

le croisement	*the cross*
le bon croisement	*the blood nick*
croiser	*to bring in new blood*
être pleine de	*to be in foal to*
le poulinage	*foaling*
la précocité	*precocity*
la saillie	*the service*
le coit	
le saut	
les conditions de saillie:	*the service conditions:*
live foal	*live foal*
payable au 1er octobre si jument pleine	*split fee*
saillie sans conditions	*straight fee*
le prix de saillie	*the nomination fee*

saillir	*to cover*
servir	
les demandes de servir	*the buying of nominations*
ramener du sang	*to buy in new blood*
le renouvellement du sang	*heterosis*
le sevrage	*weaning*
la stérilité	*sterility*
la télégonie	*telegony*
la vacuité	*barrenness*
être vide de	*to be barren to*

II Identification

The identification of the thoroughbred has a strict code, which is used world wide.

TETE	HEAD
1 *le front*	*the forehead*
2 *le chanfrein*	*the face*
3 *le naseau*	*the nostril*
4 *le bout du nez*	*the muzzle*
5 *la lèvre supérieure*	*the upper lip*
6 *la lèvre inférieure*	*the lower lip*
7 *le menton*	*the chin*
8 *l'oeil (les yeux)*	*the eye (the eyes)*
9 *l'orbite*	*the orbit*
10 *la salière*	*the supraorbital fossa*
11 *l'oreille*	*the ear*
12 *la joue*	*the cheek*

ENCOLURE	NECK
13 *la nuque*	*the poll*
14 *la crinière*	*the mane*
15 *le plat de l'encolure*	*the side of the neck*
16 *la gorge*	*the throat*
17 *la trachée*	*the windpipe/the trachea*
18 *la jugulaire*	*the jugular furrow*

CORPS	BODY
19 *le garrot*	*the withers*
20 *le dos*	*the back*
21 *le rein*	*the loins*
22 *la croupe*	*the croup/the rump*
23 *les côtes*	*the ribs*
24 *le poitrail/la poitrine*	*the breast*
25 *le flanc*	*the flank*
26 *la pointe de la hanche*	*the point of the hip*

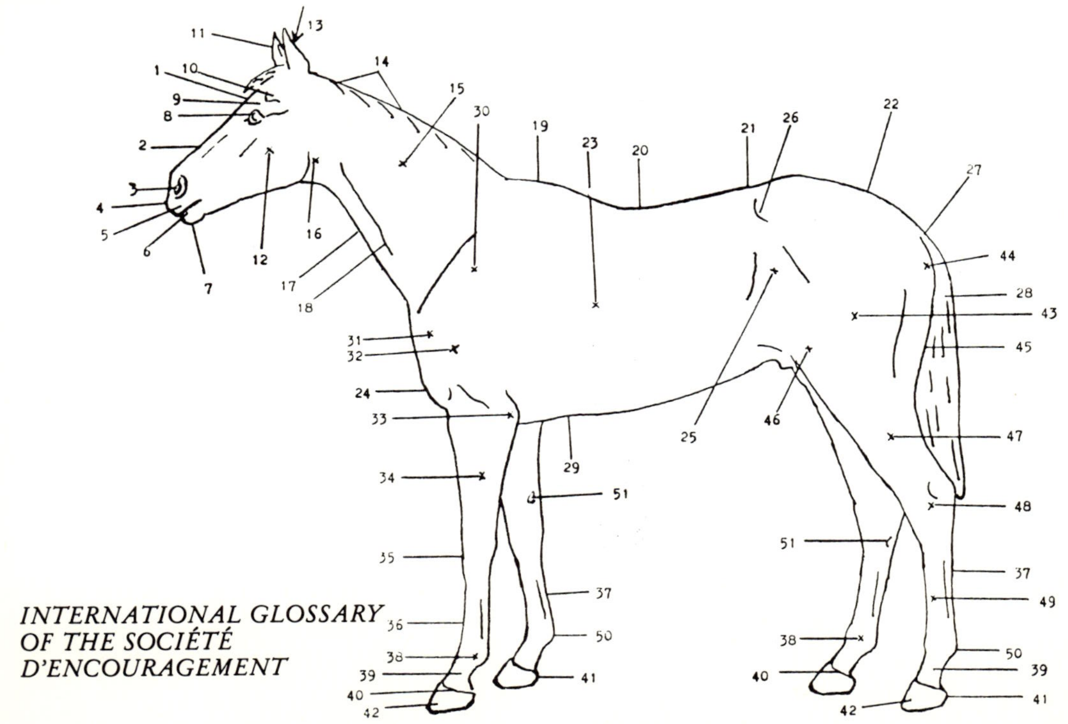
INTERNATIONAL GLOSSARY
OF THE SOCIÉTÉ
D'ENCOURAGEMENT

27 *l'attache de la queue*	*the dock*
28 *la queue*	*the tail*
29 *le passage des sangles*	*the girth*

MEMBRES ANTERIEURS — FORE LIMBS

30 *l'épaule*	*the shoulder*
31 *la pointe de l'épaule*	*the point of the shoulder*
32 *le bras*	*the arm*
33 *le coude*	*the elbow*
34 *l'avant-bras*	*the forearm*
35 *le genou*	*the knee*
36 *le canon*	*the cannon*
37 *le tendon*	*the tendon*
38 *le boulet*	*the fetlock joint*
39 *le paturon*	*the pastern*
40 *la couronne*	*the coronet*
41 *le talon*	*the bulb of the heel*
42 *le sabot*	*the hoof*

MEMBRES POSTERIEURS — HIND LIMBS

43 *la cuisse*	*the hind quarters*
44 *la pointe de la fesse*	*the point of the buttock*
45 *la fesse*	*the buttock*
46 *le grasset*	*the stifle*
47 *la jambe*	*the gaskin*
48 *le jarret*	*the hock joint*
49 *le canon postérieur*	*the hind cannon*
50 *le fanon*	*the ergot*
51 *la chataigne*	*the chestnut*

COULEUR DE LA ROBE — COAT COLOUR

alezan	*chestnut*
bai	*bay*
bai brun très foncé	*black*
bai clair	*light bay*
bai foncé	*brown*
gris	*grey*
rouan	*roan*

MARQUES	MARKS
les balzanes	*white stockings*
le coup de lance	*the prophet's thumb mark*
l'épi	*the whorl*
l'étoile/la pelote	*the star*
la liste	*the blaze/the stripe*
la raie de mulet	*the dorsal stripe*

III Commerce

SALES

The French system comprises several varieties of sale.

la vente à l'amiable	*the private sale*
la vente aux enchères	*the auction*
la vente avec redevance	*the conditional sale*

Here, an extra fee is paid to the seller above the actual sale price. Usually this sum comes from the prize money collected by the new owner.

la vente avec réserve	*the sale with a reserve price*
la vente sans réserve	*the sale without reserve price*
la vente de succession	*the executor's sale*
la vente exceptionnelle	*the special sale*
la vente spéciale	

This is a sale held at an unusual time. For instance, after a big event such as the Prix de Diane.

la vente mixte	*the mixed sale*

This comprises yearlings and other bloodstock.

la vente nulle	*the void sale*

Said of any sale which fails to satisfy legal regulations.

la vente pour dissolution d'Association	*the sale to dissolve partnership*
la vente sélectionnée	*the selected sale*

This indicates the quality of the lots for sale.

DISEASES

LES VICES RÉDHIBITOIRES

la boiterie intermittente	*lameness*
l'immobilité	*immobility*
la fluxion périodique des yeux	*moon blindness/periodic ophthalmia*
le tic	*crib biting*
l'emphysème pulmonaire	*P.E. (pulmonary emphysema)*
le cornage	*roaring*

The initials of the French list form the acronym BIFTEC — an easy way of remembering the group.

LES MALADIES CONTAGIEUSES

The ministry of Agriculture, which controls all import and export of horses, specifies that the animal must be free of the following contagious diseases:

l'anémie infectieuse équine	*E.I.A. (equine infectious anaemia)*
l'avortement à virus	*Infectious viral abortion*
la dourine	*dourine*
l'encéphalomyelite équine	*V.E.E. (Venezuelan equine encephalomyelitis)*
la gale	*mange*
la gourme	*strangles*
la grippe équine	*equine influenza*
la horsepox	*horsepox*
la lymphangite épizootique	*epizootic lymphangitis*
la lymphangite ulcéreuse	*ulcerative lymphangitis*
la métrite contagieuse	*contagious equine metritis*
la morve	*glanders*
la peste équine	*African horse sickness*
la rhinopneumonie à virus	*rhinopneumonitis*
la teigne	*tinea*

RACING DOCUMENTS

The following trace the horse's career.

la carte de saillie	*the covering certificate*
le certificat de saillie:	*the covering certificate:*
(i) la déclaration résultat de saillie	*return of mare*
(ii) le signalement du foal sous sa mère	*the foal identification certificate*
la contrat de vente:	*the sales contract:*
(i) le bordereau d'inscription	*the entry form*
(ii) les conditions de vente	*the conditions of sale*
la licence d'apprenti	*the apprentice's licence*
la licence d'entraîneur	*the trainer's licence*
la licence de jockey	*the jockey's licence*
le livret signalétique:	*the passport:*
(i) le certificat d'origine	*the registration certificate*
(ii) le feuillet de vaccination	*the vaccination certificate*
le permis d'entraîneur	*the trainer's permit*
le permis de monte	*the covering permit*
le Stud-Book français	*the French Stud Book*
the Stud-Book du trotteur français	*the Trotting Stud Book*
le Stud-Book des races françaises des chevaux de selle	*the French Saddle Horse's Stud Book*

PERSONNEL

A brief glance at "Who's who" in the French racing industry.

l'apprenti	*the apprentice*
la cavalière	*the lady rider*
le commissaire-priseur	*the auctioneer*
les commissaires	*the stewards*
le courtier	*the bloodstock agent*
le driver	*the driver*
l'éleveur:	*the breeder:*
(i) *l'éleveur-vendeur*	*the breeder-seller*
(ii) *le naisseur*	*the breeder (owner of the mare at the time that she throws her foal)*
l'entourage du cheval	*the connections of the horse (owner, trainer and jockey)*
l'entraîneur:	*the trainer:*
(i) *l'entraîneur particulier*	*the private trainer*
(ii) *l'entraîneur public*	*the public trainer*
l'étalonnier	*the stallion man*
la Fegentri	*the Federation of Gentlemen Riders*
le handicapeur	*the handicapper*
les hommes d'écurie:	*stable personnel:*
(i) *le lad*	*the lad*
(ii) *le premier garçon*	*the head lad*
(iii) *le garçon de voyage*	*the travelling head lad*
(iv) *le stud-groom*	*the stud groom*
le jockey	*the jockey*
le Jockey-Club	*the French Jockey Club*

Founded by a select group of Société d'Encouragement members in 1840. Today, it is composed of breeders and owners who promote the sport in France.

les juges: — *the judges:*
   (i) le juge à l'arrivée — *the finishing judge*
   (ii) le juge du départ — *the starting judge*
   (iii) le juge responsable de la pesée — *the clerk of the scales*

le propriétaire: — *the owner:*
   (i) la dame-propriétaire — *the lady owner*
   (ii) l'eleveur-propriétaire — *the owner-breeder*
   (iii) le propriétaire-entraîneur — *the owner-trainer*

le secrétaire de l'hippodrome — *the clerk of the course*

la Société-Mère — *the Racing Authority*

Three authorities exist in France to administer racing:

   (i) la Société d'Encouragement

for flat racing, holding its main events at Longchamp, Chantilly and Deauville.

   (ii) la Société des Steeple-Chases de France

with its centre at Auteuil for jumping events

   (iii) la Société d'Encouragement à l'Elevage du Cheval français

which is in charge of trotting at Vincennes.

le sportsman — *the racing patron*
le starter — *the starter*
le turfiste — *the punter*
le vétérinaire — *the veterinary officer / the veterinary surgeon*

IV   Racing Administration

THE RULES OF RACING

The following terms are used in the various codes of the racing authorities.

    l'accord d'exploitation limitée
This has no equivalent in the English system. It consists of a contract between the buyer and seller of a horse to stipulate that the animal will not contest a public race.

    les allocations   *the purses*
The term used to denote place money — 2nd, 3rd and 4th.

    les amendes   *fines*
The fund is the result of sums collected from fines, the registration of colours, the issuing of licences and such items.

l'appel	*the appeal*
l'association	*the partnership*
le syndicat	*syndication*
la société civile	*the public company*

Each term designates a joint way of owning a horse.

l'autorisation d'entraîner	*permission to train*
l'autorisation de faire courir	*permission to race*
l'autorisation de monter	*permission to ride*

All such authority is given by the racing body concerned.

    le Bulletin officiel   *the Official Racing Calendar*
This lists the year's events. Past races are recorded in *le calendrier officiel.*

    le certificat médical   *the medical certificate*
This confirms that a jockey is fit and must be presented with his *livret médical* or medical record book. The certificate is necessary after an interruption in his career during a season.

la cession d'engagements   *the transfer of entries*

After a sale, the right to race a horse belongs to the new owner. He takes responsibility for events in which the animal has already been entered.

le dead-heat   *the dead heat*
les dénominations   *age specifications*

According to the code, each runner conforms to one of the following descriptions:

2, 3, 4 ans:
    poulains entiers   *colts*
    hongres   *geldings*
    pouliches   *fillies*
5 ans et au-dessus:
    chevaux entiers   *horses*
    hongres   *geldings*
    juments   *mares*
disqualifié   *not eligible to race*

A description for any horse which fails to satisfy the conditions of a race.

distancé   *disqualified from the official placings*
l'écharpe   *the second colours*

To distinguish two runners belonging to the same owner, the second jockey wears a scarf, attached to the jacket.

l'engagement   *the entry*

This denotes the written application. *L'entrée* is the fee itself.

l'enquête d'office   *the enquiry*
entraîné en province   *trained in the provinces*

This indicates that the horse was trained at a centre other than Chantilly, Maisons-Lafitte or Deauville.

le forfait   *the forfeit*

The French forfeit list is known as the *Liste des Oppositions*.

le formulaire

This is a general term for the conditions of an event:

    nés et élevés en France   *born and bred in France*
    pour tous chevaux   *open to all horses*
    pour tous poulains et   *open to all horses up to*
    pouliches   *4 years*

On the race card, the runner is denoted as:

M.	for colts and geldings
H.	for geldings
F.	for fillies and mares

le non-paiement   *arrears*

Designates any fee owing to the racing authority.

le pari particulier   *the private bet*

A description for any private wager between two owners.

les primes aux éleveurs   *breeders' premiums*

A breeder is eligible to receive:

15% of the prize money if his horse is amongst the first four runners

25% of the sum when this happens in a pattern race

10% of the purse if the horse runs first or second in an event outside France.

les primes aux propriétaires   *owners' premiums*

The owner may collect:

30% of the purse when the horse runs amongst the first four

50% of the prize money when this occurs at Longchamp, Deauville or Chantilly.

le prix   *value to the winner*
le valeur nominale
le montant d'un prix
la réclamation   *the objection*
déposer une réclamation   *to lodge an objection*
la Région parisienne   *Paris area regulations*

This term refers to the qualification and weights applied to horses which have run or won on Paris courses: Longchamp, Chantilly, Deauville, Saint-Cloud, Maisons-Lafitte, Le Tremblay, Vichy and Evry. This regulation applies to flat racing only.

rétrogradé   *relegated*
le walk-over   *the walk over*

RACING COLOURS

There are 18 colours for the silks:

beige	*cream*
blanc	*white*
bleu	*blue*
bleu-clair	*light blue*
bleu foncé	*dark blue*
gros bleu	
grenat	*maroon*
jaune	*yellow*
marron	*brown*
mauve	*mauve*
noir	*black*
orange	*orange*
rose	*pink*
vert	*green*
vert-clair	*light green*
vert foncé	*dark green*
gros vert	
violet	*purple*

Motifs for the jacket, sleeves and cap can be studied on the accompanying diagrams.

DISPOSITIFS DE COULEURS
RACING COLOUR SCHEMES

I *CASAQUES*: BODY

1 *Unie*
Plain

2 *Bande*
Strip

3 *Bretelles*
Braces

4 *Ceinture*
Belt

5 *Cerclée*
Hooped

6 *Chevron*
Chevron

7 *Chevrons*
Chevrons

8 *Coutures*
Seams

9 *Croix de St-André*
Cross Belts

10 *Croix de Lorraine*
Lorraine Cross

11 *Damiers*
Check Squares

12 *Losange*
Diamond

13 *Losanges*
Diamonds

14 *Ecartelée*
Quartered

15 *Epaulettes*
Shoulders

16 *Etoile*
Star

17 *Etoiles*
Stars

18 *Pois*
Spots

19 *Rayée*
Stripes

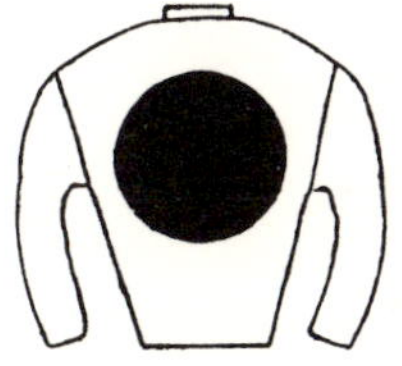

20 *Disque*
Disc

DISPOSITIFS DE COULEURS
RACING COLOUR SCHEMES

I *MANCHES*: SLEEVES

1 *Unies*
Plain

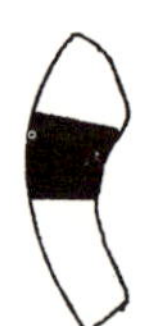

2 *Brassards*
Armlet

3 *Cerclées*
Hooped

4 *Coutures*
Seams

5 *Etoiles*
Stars

6 *Pois*
Spots

7 *Rayées*
Striped

8 *Chevrons*
Chevrons

9 *Damiers*
Check Squares

10 *Losanges*
Diamonds

III *TOQUES*: CAP

1 *Unie*
Plain

2 *Cerclée*
Hooped

3 *Rayée*
Stripes

4 *Damiers*
Check Squares

5 *Pois*
Spots

6 *Ecartelée*
Quartered

7 *Etoiles*
Stars

8 *Losanges*
Diamonds

RACE CATEGORIES

La course à conditions *the condition race*

This describes any race subject to special regulations concerning the runners, such as age or weight.

la course à poids pour âge *the weight for age event*

la course attelée *the harness race*

The usual name for a trotting event.

la course classique *the classic race*

The French classic programme includes 17 flat races, 6 obstacle races and 7 trotting events.

la course de catégorie *the category race*

This term denotes races other than pattern events. There are 4 categories divided according to the prize money.

la course dédoublée   *the divided race*
Any race run in 2 divisions due to the number of runners.
la course à obstacles   *the jumping race*
This is a general term which covers:
    la course de haies   *the hurdle race*
    le steeple-chase   *the steeple-chase*
The latter is run over 3000 metres and the runners encounter tougher obstacles.
la course de maiden   *the maiden race*
On the race card, this event is described as a race *pour chevaux n'ayant jamais gagné.*
la course d'inédits   *the novices' race*
An event for 2 year olds.
la course en partie liée   *the series race*
This is a trotting event which has 2, 3 or 4 heats. If the same horse wins 2 heats, it is the victor.
la course principale   *the pattern race*
These are divided into 3 groups according to the purses.
le course publique   *the open race*
In France, this denotes any race for which the prize money exceeds 300 francs.
la course à réclamer   *the selling race*
Races where only certain runners are for sale are known as *les prix mixtes.*
le handicap   *the handicap*
    le handicap dédoublé   *the divided handicap*
    le handicap libre   *the free handicap*
This is open to all horses which qualify for the weight conditions.
le handicap limité   *the limited handicap*
For such an event, the weights are fixed in advance.
le handicap de catégorie   *the handicap according to rating*
According to their rating by the handicapper, horses can enter races suited to their weight.
le match   *the match*
A race between two cracks.

THE ORGANISATION OF A RACE

Since this procedure is very similar in all countries, the terminology will be familiar.

le champ	*the field*
les compétiteurs	
le changement de ligne	*crossing*
le corde	*the rails*
le tirage des places à la corde	*the barrier draw*
déclarer forfait pour un cheval	*to declare forfeit*
le défilé	*the parade*

A parade is held before important events such as the Prix du Jockey-Club or the Prix de l'Arc de Triomphe.

l'élimination	*elimination balloting out*
engager un cheval	*to enter a horse*
le film contrôle	*the film patrol*
les gains	*winnings*

In flat racing and hurdle events, this term denotes only the sum collected by the winner. However, in the trotting code, it means any purse from first to fourth position.

la monte	*mounts*

This general term covers the regulations relating to the engagement of a jockey.

le changement de monte	*change of rider*
la déclaration de monte	*declaration of riders*
le tarif de monte	*riding fee*
l'équipement de monte:	*riding gear:*
le bonnet	*the hood*
la bride	*the bridle*
la casque de protection	*the skull cap*
le collier de chasse	*the breastplate*
le filet	*the snaffle bit*
les guêtres	*the boots*

la martingale	*the martingale*
la muserolle	*the noseband*
les oeillères	*the blinkers*
la selle	*the saddle*
la serviette numérotée	*the number cloth*
le sulky	*the sulky*
le tapis de selle	*the weight cloth*

N.P.

Two terms are represented by these letters:

non partant	*scratched*
non placé	*unplaced*
les objets d'art	*trophies*
le parcours	*the journey*
	the run

The code uses this term to refer to the race, which has several important stages:

le départ:	*the start:*
le départ à l'australienne	*the starting gate*
le départ aux élastiques	*the tape start*
le départ au drapeau	*the flag start*
le départ en stalles	*the start from the stalls*
l'autostart	*the mobile start*

This term is used in trotting.

l'essai de départ en stalles	*the trial start*
etre sous les ordres du juge du départ	*to be under the starter's orders*
le faux départ	*the false start*

The following terms figure in the description of a race:

la montée	*the rise*

Said of the undulation in the track at Longchamp.

la ligne d'en face	*the back straight*
à mi-parcours	*at the half-way mark*
la descente	*the descent*

Again this is a reference to Longchamp.

le dernier tourant	*the last corner*
la distance	*the distance (200 metres from the post)*

dans la ligne d'arrivée   in the straight
l'arrivée   the finish
  le classement d'arrivée   the finishing order
  le poteau d'arrivée   the post
  les distances d'arrivée   the margins
    le dead-heat   dead heat
    un nez   a nose
    une courte tête   a short head
    une tête   a head
    une courte encolure   a short neck
    une encloure   a neck
    une demi-longueur   half a length
    une longueur   a length
    loin   distant
  le tour   the circuit of the track
  les partants   the starters
         the runners

la déclaration des partants   the list of runners
la piste: the track:
  la piste à main droite   the right handed track
  la piste à main gauche   the left handed track
  la ligne droite   the straight course
This is denoted by the letters L.D. on the race card.
  la piste d'entraînement   the training track
  la piste de gazon   the grass track
  la piste en sable   the sand track
la pelouse   the public enclosure
At Longchamp, the lawn enclosure is known as *le pesage*.
la photo-finish   the photo
le poids   the weight
A series of terms is connected to this important concept:
  la décharge   the allowance
  la remise de poids
  la surcharge   the penalty
  la barème de décharges   the scale of allowances
The Code fixes these allowances for jockeys and apprentices.
  l'échelle de poids   the weights
Said of the range of weights in a particular event.

la publication de poids	*the publication of the weights*
recevoir du poids	*to carry less*
rendre du poids	*to carry over*
la salle de balances	*the weighing room*
le pesage	*the weigh-in*
la pesée avant la course	*the weigh-out*
la pesée après la course	*the weigh-in*
rentrer aux balances	*to return to scale*
le top-weight	*the top-weight*
le bottom-weight	*the bottom weight*
le programme	*the race card*
représentant	*representing the owner*

This term is designated on the race card by an R, placed next to the name of the owner. This indicates that the real owner has temporarily relinquished his right to race the horse to a representative. His usual colours are used but his name does not appear on the programme. This practice is also followed in cases of bereavement.

rester au poteau (R.P.)	*left at the post*
la réunion	*the meeting*
le meeting	*the one day meeting*
la diurne	*the daylight meeting*
la nocturne	*the night meeting*

Both these terms refer to trotting fixtures, which usually take place in the evening.

le rond de présentation	*the ring*

Said of the parade ring in the lawn enclosure at Longchamp.

le temps	*the official time*
la tenue de course:	*racing attire:*
les bottes à revers de jockey	*jockey boots*
la casaque	*the jacket*
la cravache	*the whip*
la culotte blanche	*the riding breeches*
les étriers	*the stirrup irons*
la toque	*the cap*

THE GOING

The ground is officially described by one of the following terms:

sec	*hard*
ferme	*firm*
bon	*good*
assez souple	*dead*
souple	*yielding*
collant	*soft*
très souple	*holding*
lourd	*heavy*

THE BETTING SYSTEM

All operations are controlled by *Pari Mutuel Urbain (P.M.U.)*, which has depots throughout France.

On course betting is the responsibility of *Pari Mutuel Hippodrome (P.M.H.)*

A variety of bets is available:

le pari simple gagnant	*the win bet*
le pari simple placé	*the place bet*
le pari couplé	

This bet is available off course only. Two horses are selected in the same race. The bet is taken as either:

gagnant

where the horses must arrive first and second, or

placé

where they must finish amongst the first 3 runners.

le pari super-couplé

This is another off course bet: 3 runners are chosen; if 2 arrive first and second, a dividend is paid. If all 3 are placed, the return is much greater.

le pari jumelé

This is the *couplé* bet, but available on course.

le pari "par reports"

A series of win or place bets is taken on races at the same meeting. This is available off course at *P.M.U.* and is similar to the Accumulator in Britain.

le pari tiercé
Three horses are selected and the dividend is paid according to the order designated. The bettor specifies whether he is taking the wager:

dans l'ordre exact	*in correct order*
dans l'ordre inexact	*in any order at the post*

The *tiercé* is an off course bet.

le pari triplet
This is the on course equivalent of the *tiercé*.

le pari trio
Another on course triple bet. The bettor chooses 3 horses to occupy the first 3 places.

la pari quarté
A 4-horse bet which is available:

dans l'ordre exact	*in correct order*
dans l'ordre inexact	*in any order at the post*
faire écurie	*bracketed*

If the same owner has entered 2 runners in the same event, *Pari Mutuel* may bracket them in the betting. Thus, win tickets are valid on either runner.

champ total	*the field*

In a *couplé* or *jumelé* bet, one horse may be bracketed with each of the other entrants.

champ réduit	*selected field*

This term denotes the bracketing of one horse with a limited selection from the other competitors.

V Performance

THE GAITS

The terms relating to movement are:

le canter	*the canter*
	the canter down
le canter à demi-train	*the slow canter*
le galop	*the gallop*
le pas	*the walk*
le trot	*the trot*
l'amble	*pacing*

Pacing is forbidden in France so this term is used to denote the movement permitted in the United States and Australasia.

TRAINING TERMS

This is a selection of some of the best known expressions:

l'action	*movement*
avoir une action étendue	*to stride out*
avoir une belle action	*to move well*
avoir une mauvaise action	*to move badly*
avoir une vilaine action	
être affûté	*to be ready*
être au top	
être prêt	
affûter un cheval	*to prepare a horse*
amener un cheval à son meilleur niveau	
amener un cheval à son mieux	
les aplombs	*the stance*
être apte	*to be suitable for training*

l'apparence	appearance
avoir une bonne apparence	to look well
avoir de l'allure	to be good looking
les aptitudes	ability
avoir l'aptitude à la course	to have racing ability
avoir l'aptitude à la distance	to be suited by the distance
avoir l'aptitude au mile	to get the mile
avoir l'aptitude au terrain	to like the going
l'arrière-main	the hind quarters
l'atteinte	brushing
l'avant-main	the front
les bandages	bandages
etre capable de progresser	to have scope
le caractère	character
le tempérament	temperament

While these can be equated in English, the French terms have a positive or negative connotation. This will be evident from the following series:

être allant	to be spirited
avoir un bon tempérament	to have a good temperament
être docile	to be placid
être maniable	to be easy to handle
être brave	to be brave
être courageux	to be game
être honnête	to be genuine
être régulier	
avoir du caractère	to be difficult
être difficile	
être méchant	
être rétif	
le cheval de distance moyenne	the middle distance horse

Said of the horse which prefers to race over 8 to 12 furlongs.

le cheval de grande classe	*the class horse*
le cheval de premier ordre	

This describes the ability of the runner.

la conformation	*conformation*
le modèle	

Many terms are associated with this concept:

être développé	*to be well-grown*
être décousu	*to be badly made*
être équilibré	*to be balanced*
être fort	*to be strong*
être harmonieux	*to be well-made*
être léger	*to be light-framed*
être peinture anglaise	*to be lean*
être racing-like	*to be whippety*
être près de terre	*to be neat*
	to be compact
être proportionné	*to be well-proportioned*
être racé	*to be true to type*
être uni	*to be close coupled*
le débourrage	*breaking-in*
être débourré	*to be broken-in*
le défaut	*the defect*

A general term to denote any fault.

être doué	*to be talented*
avoir de l'étoffe	
être promis à un bon avenir	
l'entraînement	*training*
arrêter l'entraînement	*to let down*
sortir de l'entraînement	*to retire*
retirer de l'entraînement	*to retire for a spell*
reprendre l'entraînement	*to resume work*
être surentraîné	*to be over the top*
l'état	*form*
la forme	
la baisse de forme	*the loss of form*
être hors de forme	*to be unfit*
être passé de forme	
être en état	*to be fit*

être en grande forme
être en pleine forme
être fit
les ferrures   racing plates
le feu   firing
le flyer   the short runner

This designates the horse which does well over about 5 furlongs.

le fourrage: fodder:
     l'avoine   oats
     le foin   hay
     le fourrage sec   roughage
     le mash   mash
     la ration journalière   the daily ration
     la ration de travail   the working ration
     le vert   green fodder
l'inédit   the novice
être limité   to be of limited ability
un second plan
la machine à galoper   the racing machine

A familiar term for any fast horse.

les maladies du pied: diseases of the foot:
     le crapaud   canker
     la crevasse   cracked heels
     la fourbure aiguë   laminitis
          founder
     la maladie naviculaire   navicular disease
     les molettes   wind galls
     la pourriture de la   thrush
     fourchette
     la seime   sand crack
la maréchalerie   farriery
le miler   the mile specialist
le cheval de 1600 mètres
les ordres   orders
     monter aux ordres   to ride to orders
être précoce   to be precocious
     to be an early type

la précocité	*early development*
la qualité	*breeding*
être bien né	*to be well bred*
être de belle origine	
le sprinter	*the sprinter*

Said of the good performer over 5 to 6 furlongs.

les tares de membres:	*limb faults:*
le claquage	*the sprain*
être claqué	*to be broken down*
l'éparvin	*bone spavin*
la forme	*ring bone*
la jarde	*curb*
le suros	*splint*
le vessigon	*bog-spavin*
la tenue	*staying ability*
le fond	
le cheval de tenue	*the stayer*
le cheval de fond	
le stayer	

Associated concepts are:

la résistance	*stamina*
la trempe	
être résistant	*to be tough*
être bien trempé	
tenir la distance	*to get the trip*
le travail	*work*
résister au travail	*to take to training*
être à court de travail	*to be short of work*

THE TRACK RECORD

The range of terms pertaining to this area is endless — such is the variety and inventiveness of the racing press. However, this list assembles some of the best known and most frequently used expressions.

l'actif	*the win record*

Said of the horse's victories during a season or for his career.

l'allure	*the pace*
le train	
mener à grande allure	*to force the pace*
forcer l'allure	
faire le train	*to set the pace*
assurer le train	
courir à bon train	*to bowl along*
mener grand train	*to set a cracking pace*
courir à un train régulier	*to run at a steady pace*
une course sans train	*a race without genuine pace*
être à l'ouvrage	*to be under pressure*
s'arrêter	*to pull up*
attendre	*to lie up handy*
faire une course d'attente	
l'avantage	*the advantage*
avoir l'avantage	*to have the advantage*
être en tête	
prendre un léger avantage	*to have a narrow lead*
prendre une tête	
prendre un net avantage	*to have a clear advantage*
s'assurer l'avantage	*to take the lead*
prendre le commandement	
prendre la tête	
relayer en tête	
garder l'avantage	*to keep the advantage*
conserver le meilleur	
dominer	

avoir le mors aux dents   *to run on*
être bien battu   *to be well beaten*
être bouscoulé   *to receive interference*
être gêné
Said of the horse which is knocked or blocked.
la campagne   *the campaign*
Refers to the runner's programme for a particular season.
causer des déceptions   *to disappoint*
décevoir
âtre décevant
céder   *to give up*
le cheval de jeu   *the pacemaker*
  faire le jeu   *to set the pace*
combler son retard   *to make up lost ground*
refaire du terrain
venir de loin
le compagnon de box   *the stable companion*
le compagnon d'écurie
couper la ligne   *to cross*
courir   *to run*
  courir bien   *to run well*
  courir d'une façon
  prometteuse
  se distinguer
  courir mal   *to run badly*
  courir obscurément   *to run poorly*
le crack   *the crack*
  le crack jockey   *the crack jockey*
le déboulé   *the fast, close finish*
le Derby-winner   *the Derby winner*
This is a press term.
la déroute   *the defeat of the favourite*
                                 the upset
se détacher   *to go clear*
être distancé   *to be disqualified*
Said of the horse which loses its official place.
le doublé   *the double*
Denotes a double victory in 2 important events.
l'effort   *effort*

faire un effort	to challenge
passer à l'offensive	
fournir un bel effort	to make a genuine effort
prononcer son effort	to improve
être en détresse	to be in distress
être en queue	to bring up the rear
fermer la marche	to bring up the van
occuper le dernier rang	
photographier la queue	to tail the field
ramasser les casquettes	
l'épreuve	the race
la tentative	
la super-épreuve	the top event
le top-event	
l'essai	the trial
s'étendre	to quicken
faiblir	to weaken
faiblir à la distance	to weaken in the straight
faire un canter	to win easily
filer en tête	to be a front runner
galoper aux premiers rangs	
la fin de course	the finish
le finish	
la lutte	
la rush	
une arrivée disputée	a close finish
une arrivée serrée	
finir en trombe	to finish fast
finir vite	
finir fort	to finish well
fournir une bonne fin de course	
être froid	to fail to respond

Said of the horse which does not reply to his jockey's request for a final effort.

le gagnant	the winner
le lauréat	
le vainqueur	

inquiéter	*to threaten*
menacer	
le lâché	*the good lightweight*
le leader	*the leader*
le lot	*the field*
le pelotan	
se mettre sur ses jambes	*to get into stride*
manifester de la névrosité	*to be edgy*
l'outsider	*the outsider*
partir lentement	*to dwell*
être stiff	
partir vite	*to start well*
	to get a flyer
le passage	*the run*
avoir un bon passage	*to have a good run*
avoir un mauvais passage	*to have a bad run*
forcer un passage	*to find an opening*
passer à l'extérieur	*to go round the outside*
passer sans effort au	*to stride away*
commandement	
la place	*the placing*
l'accessit	
le poteau	*the winning post*
franchir le poteau	*to cross the line*
rallier le poteau	*to go for the post*
rester au poteau	*to be left at the post*

This is denoted by the letters R.P. on the card.

prouver sa classe	*to prove one's class*
donner la mesure de sa	
supériorité	
le record de piste	*the track record*
le détenteur d'un record	*the record holder*
la détentrice d'un record	
reprendre un cheval	*to check a horse*

Often the reason for a protest.

répondre aux sollicitations	*to respond*
du jockey	*to come alight*
la rentrée	*the return*

faire sa rentrée	*to resume*
résister	*to hold on*
résister au rush	*to hold on at the finish*
résister à l'attaque	*to keep going*
le retour en forme	*the return to form*
la réussite	*the victory*
le succès	
la victoire	
confirmer un succès	*to win again*
rééditer un succès	*to follow up*
serrer la corde	*to race on the rails*
la sortie	*the outing*
être surclassé	*to be out of one's class*
tirer	*to pull*
trouver sa forme	*to find form*